This book belongs to

240 Madison Avenue
Yuma, Arizona

Designed by P. R. Callahan
C. C. Reeves

FIRST EDITION

Rube, B. Johnny, 1937—
A wooden road through the hollow of God's hand : taming the American Sahara / by B. Johnny Rube.
p. cm.
Includes bibliographical references (p. 65 - 69).
ISBN 1-884172-14-8 (alk. paper)
1. Algodones Dunes (Calif.) — Description and travel. 2. Roads, Plank — California — Algodones Dunes — History. I. Yuma County Historical Society (Yuma County, Ariz.) II. Title.
979.1'71—dc21 96-48295
CIP

Printed in Arizona U.S.A.
by
Sun Graphics

A Wooden Road Through The Hollow of God's Hand

Taming The American Sahara

By B. Johnny Rube

Dedicated
To The
Memory and Spirit of Ed Boyd

Cover - Second Plank Road through the Algodones Dunes

Preceeding Pages iii, iv, v - Dune Buckwheat ***Eriogonum deseaticola*** *Algodones Dunes July 1996*

Overleaf Dune Buckwheat Algodones Dunes July 1996

Overleaf Right Creosote Bush is the most characteristic species of the hot North America deserts. Called the "desert drugstore" by native people for its many medicinal uses, this evergreen shrub exudes a pungent aroma after desert rain storms.

Page ix and Overleaf

Creosote Bushes, moderate size shrubs, three to four feet high have adapted to the shifting sands through rapid growth or gigantism. Although it cannot survive on the active dunes of the dune field, creosote flourishes to heights of five to six feet on the stable dunes on the edge of the field. When threatened by sand accumulating around the young plants, the main stem elongates, keeping the tips of the branches above the moving sand. If the sand moves away from the plant, the elongated stems lose support and topple over from the weight of the leafy tips forming an arch.

Overleaf right
Algondones Dunes July 1996

page xiii
Algodones Dunes July 1996

CONTENTS

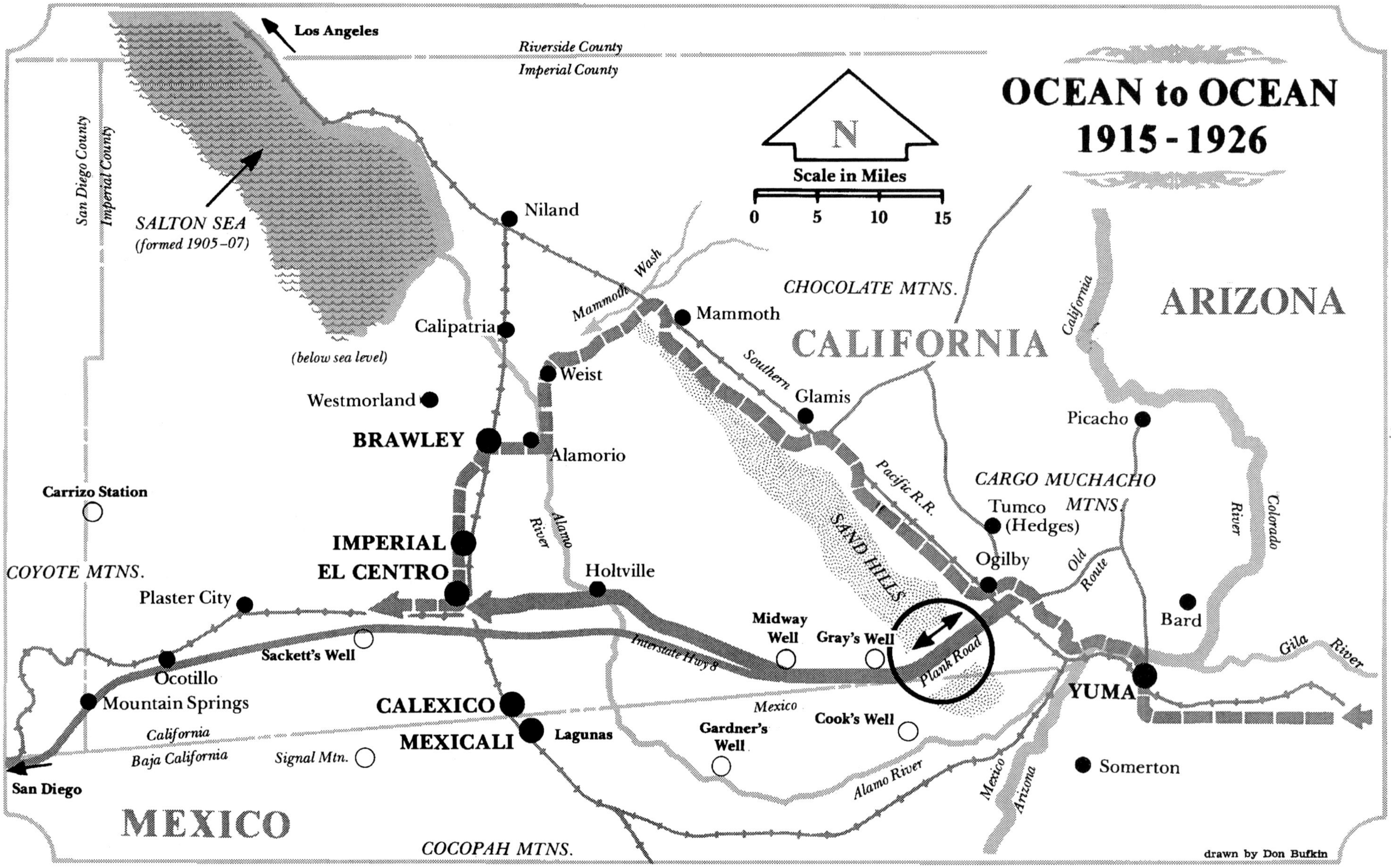
OCEAN to OCEAN
1915 - 1926
Los Angeles
Riverside County
Imperial County
San Diego County
Imperial County
SALTON SEA
(formed 1905–07)
(below sea level)
N
Scale in Miles
0 5 10 15
Niland
Calipatria
Westmorland
Weist
BRAWLEY
Alamorio
Mammoth Wash
Mammoth
CHOCOLATE MTNS.
CALIFORNIA
ARIZONA
Southern Pacific R.R.
Glamis
SAND HILLS
CARGO MUCHACHO MTNS.
Tumco (Hedges)
Ogilby
Old Route
Picacho
California
Arizona
Colorado River
Bard
Gila River
Carrizo Station
COYOTE MTNS.
IMPERIAL
EL CENTRO
Alamo River
Holtville
Plaster City
Ocotillo
Mountain Springs
Sackett's Well
Midway Well
Gray's Well
Plank Road
Interstate Hwy 8
CALEXICO
MEXICALI
Lagunas
California
Baja California
Signal Mtn.
San Diego
Mexico
Gardner's Well
Cook's Well
Alamo River
Mexico
Arizona
YUMA
Somerton
MEXICO
COCOPAH MTNS.
drawn by Don Bufkin

INTRODUCTION

Get Out and Push

Today a modern divided highway crosses the southeastern end of Imperial Valley, but not so long ago traveling across the dunes was a problem. Anyone who has ever crossed the dunes in an automobile, when there was a breeze of any magnitude, will remember the eerie feeling produced by the sight of the drifting, swirling sands. It appears as though one is traveling crosswise in a running stream of water. Nature also occasionally reminds travelers of its power when strong winds lash Interstate 8 with sand that can strip the paint off a vehicle in no time at all.

The Algodones Dunes posed an insurmountable barrier for early travelers, but men of vision came along and built a wooden road across the dunes. Primitive it was, but it served the needs of early motoring traffic until a better road could be built. In the good ol' days, journeys over the dunes were something of a hazard as the road was vulnerable to the sifting sand. During sudden storms, sand drifting across the road completely buried it. If the motorist carried a shovel, he would clear the sand and continue on. With the completion of the pavement in 1926, the old plank road was abandoned. As the years rolled by, weather and man took a heavy toll of the old planks and little evidence of its existence remains.

While gathering materials about the building of the plank roads across the dunes, it became quite evident that there wasn't a single book published about this subject. There have been many short articles written, but due to their brevity, they often leave the reader wanting to know more information. Probably two of the best such short stories are: "The Plank Road" by James Bates and "From Dusty Trails to Superhighways" by James D. Blick. I have endeavored to present the entire story of the plank road and the problems encountered in its construction. For further reading on this subject, the reader may refer to the bibliography.

To those authors and publishers who gave me permission to borrow excerpts from their material, I am deeply grateful. A very special thanks to Barbara Christensen and the late Harvey Johnson for their kind assistance. My deep felt appreciation to a friend, Lori Davidson of Tucson whose inspiration helped me across the rough spots and gave me confidence to research and write about Yuma's history. And then there are two ladies, a mother and daughter duo without whose assistance this book would not be in print. To Pat Callahan and Cathy Reeves goes a heartfelt "Thank You Ladies." Lastly, any mistakes or shortcomings in these pages are mine alone.

B. Johnny Rube
Yuma, Arizona
1996

The Algondones Dunes

THE ALGODONES DUNES

How many times have travelers crossed the Algodones Dunes and wondered how they came to be there? The Mexicans know the dunes as "La Palma de la Mano de Dios(1)* The Hollow of God's Hand." It's also variously known as the American Sahara, Algodones Dunes and to the local folks - the Sand Dunes.

The Algodones Dunes, which are located in the great Colorado Desert, (2)* form a great sand tract stretching more than 40 miles along the eastern margin of the Imperial Valley in Southern California. (3)* They lie in a northwesterly direction, three to six miles wide and some dunes are 200 to 300 feet high.

They are located in the Cahuilla Basin, a structural depression much of which is below sea level, and includes the Imperial and Coachella Valleys. The lowest portion of the basin is now the Salton Sea. (4)*

A large lake occupied Cahuilla depression until perhaps 300 years ago; (5)* however, it is not certain whether it filled more than once. If the southern shore of the ancient lake lay a few miles off the international boundary as suspected, the maximum length of the lake was about a hundred miles with a width of 35 miles along a east-west line passing through the present town of Calipatria. The lake had a depth of about 300 feet. (6)*

The dunes were probably produced by wind which blew sand inshore from the beaches of Lake Cahuilla. The large volume of material in these ancient beaches indicates vigorous wave and current action along the northwestern shore of the lake. That part of the shore line between Flowing Well siding and Mammoth Wash is believed to be the main source of sand in the Algodones Dunes. (8)*

Both Robert M. and Kenneth S. Norris believe that the source of sand now in the Algodones Dunes was: (1) the highest Lake Cahuilla beach deposits, (2) sand from Mammoth Wash and associated drainages, (3) lesser amounts of sand from East Mesa, and (4) sand from the Chocolate Mountain drainages south of Mammoth Wash. (9)*

Apparently northwesterly winds began to move the sand on shore from Lake Cahuilla beaches immediately following development of the highest lake level. This sand quickly developed into a chain of closely spaced dunes, advancing across the desert floor to the southeast. (10)*

The yellow sands (11)* are of granite origin (12)* and are derived not only from the disintegrating of the rocks of the mountains bounding the great Salton depression, but also from the breakdown of the Old Tertiary marine beds. (13)* It is unknown how old the dunes are, but probably of considerable age.

Overleaf - Dune Buckwheat
Algodones Dunes July 1996

Hummocks are formed by sand blowing into the leaves and jumbled stems of dunes plants. Big galleta grass HILARIA RIGIDIA builds hummocks, tangled mats of stems, roots and new growth so tightly bound that high winds cannot destroy them.

EARLY TRAVEL ON THE DUNES

The dunes posed a formidable obstacle to the early travelers. Juan Batista de Anza in 1774, was unable to cross the dunes when his guide Sebastian Tarabal was not able to recognize the trail he had blazed in 1773. They finally made it by traveling off the dunes, swinging west and then north into California. (1) During the pioneer days a great deal of freighting between San Diego and Fort Yuma was handled with pack animals. Practically all wagon hauling was either to the north or south of the dunes. (2)*

To avoid the sand hills, the Butterfield Overland Mail route crossed into Mexico, followed a line of wells and then recrossed into the United States a short distance west of what is now Calexico. (3)*

As early settlers began to enter the Imperial Valley, roads were needed to bring in supplies to the ever increasing number of farmers. (4)* There then arrived upon the scene a marvel of the modern age, the self propelled wagon: the automobile, which needed even better roads.

In the first decade of the 20th Century the automobile's popularity began to rise. The early vehicles were expensive and only the rich could afford them. However, in 1905, Henry Ford began making affordable cars. Their numbers grew rapidly and by 1910

Road construction relied on teams and scrapers.

registration, California had 44,120 vehicles. (5)* As soon as cars became available, groups of men formed auto clubs. One such club was the Automobile Club of Southern California founded in 1910. (6)* This organization will play a large part in the plank road drama.

Before 1907, what is now Imperial County, was part of San Diego County. The five towns of Calexico, Holtville, Brawley, Imperial and El Centro were growing fast and roads were needed so farmers could ship produce and goods to markets on both coasts. By 1910 much of the emphasis in California and especially San Diego was on good roads to Imperial Valley and to the distant east. San Diego needed an outlet to the east which would be shorter than going through Los Angeles. (7)*

It was evident that usable roads in and out of Imperial Valley were needed. Many felt that the road should go to San Diego and become a part of a transcontinental highway from San Diego and Imperial Valley. (8)*.

It had been San Diego County Surveyor S. L. Ward who in 1901 (10)* warned San Diego that they had better look to its future relations with the people of Imperial Valley, as Los Angeles merchants were attracted to that region. He further said that unless a better wagon road was constructed, the trade of the prospering valley would go to Los Angeles. The promise to build the Panama Canal had a profound influence on road building in California. The building of the transcontinental railroads left San Diego on a branch line while Los Angeles received two great terminals. San Diego was isolated in the southwestern tip of the United States with Mexico to the south and high mountains as well as the Colorado Desert to the east. Good roads and railroads east towards Yuma seemed to be the only answer.

By 1910 the first step was taken for an auto road to Imperial Valley. (11)* Los Angeles decided it was being threatened and hastened to work on a route to the valley by way of San Bernardino and Mecca. San Diegans then sought contributions from concerned citizens and raised $60,000 for a new road from San Diego County line to the desert floor in Imperial Valley.

Also in 1910, the Ocean to Ocean Highway Association met in Phoenix with 84 delegates from California, Arizona, and New Mexico. (12)* Arizona and New Mexico had no trouble agreeing on a route to Yuma. Los Angeles wanted to go by way of Salton Sea which would have removed the Imperial Valley's principal cities from the proposed road system.

Yuma wasn't going to be left out of the road question. On October 10, 1911, Amos Adams and A. B. Ming, both of Yuma, loaded a wagon with provisions in order to "spy out" an All American Road from Yuma to California. (13)*

They returned eight days later and filed their mileage stating that the continental highway should be built via the southern route. Their travels led them to Ogilby, Glamis, Mammoth Wash to Brawley and hencé El Centro. (14)* In Brawley they met with Editor M. D. Witter of the Brawley News and Imperial County Supervisor R. E. Willis. Both men were very receptive to this idea and promised to work with the Yuma people for "better roads".

Preroad travel through the dunes tested the mettle of the traveler.

Now came upon the Imperial Valley scene, a man who would be most instrumental in building the road across the valley and the Algodones Dunes. He was Col. Ed Fletcher, an ardent booster of the National Highway and the Yuma-San Diego route. Fletcher felt that was the most practical way to reach the Pacific from the Atlantic Coast. (1)*

POLITICS AND ROAD BUILDING

An auto race was to decide whether the San Diego or Los Angeles route was more practical. On the San Diego team was Fletcher, F. B. Naylor, and Fred W. Jackson. A prize of $3,000 was raised and Phoenix citizens chipped in another $1,000. The San Diego entry became the **Tribune-Gazette** auto and was driven by Fletcher. The **Los Angeles Examiner** driver was given the privilege of selecting any route that he wished. His decision was to race via Blythe to Phoenix.

Fletcher drove his four-cylinder air cooled Franklin through El Centro and toward the distant hills. He took no chances and stationed six horses in the dunes. When his car bogged down, the team pulled him out. At Yuma he crossed the Colorado River on the Southern pacific Railroad bridge at night and arrived in Phoenix 19½ hours later. The Los Angeles entry broke down and never finished. (2)* Fletcher was hooked and became an enthusiastic booster for the road across the sand hills.

In his memoirs, Fletcher had this to say about the sand hills and the National Highway in particular: "The sand hills between Holtville and Yuma forced all cars to go 46 miles further around by the way of Mammoth Wash and Niland, then follow the Southern Pacific Railroad to Yuma. With sand eight to ten inches deep, it was a grueling test as well as dangerous. Los Angeles papers urged construction of a National Highway via San Bernardino and Mecca to Yuma and then east. San Diego was at the end of the trail and tourists had to come in and out via the coast from Los Angeles to San Diego and back."

Pumping up tires. Heading for Mammouth 5 miles.

On the way to Yuma.

Arriving at Brawley Garage.

Louis in some very deep sand.

*At the mouth of Mammouth Wash.
The worst sand wash in the country.*

On the way after quite a little rest.

*Going down a steep bank to a ferry
boat at Yuma.*

Leaving Yuma, very deep dust.

Nineteen hundred and twelve was a very important year for the sand dunes as well as the beginning of problems that "surfaced" in placing a solid road across the dunes. The first try at crossing the sand hills was made by Imperial County Supervisor Edwin E. Boyd from Holtville District, who experimented with a brush mat road at a cost of $3,500. The first few heavy cars that traveled over the road ground it into bits. A cry went up from taxpayers of "wasted money". The county was young and poor and that was a lot of money, but there had to be a start. Give Boyd credit for being a pioneer: he started something and will be heard from again in the road saga. (3)*

In April 1912, Fred Fuller tried to cross the desert between Holtville and Yuma to blaze a trail for the actual Ocean to Ocean Highway. He failed to accomplish his purpose, but the San Diego people experimented by sending a car through Devils Canyon to El Centro making the trip in seven hours and 50 minutes. (4)*

Yuma wasn't just sitting by, for in May 1912, a branch of the Highway Association was formed and Yumans donated $1,000 for work on the California end of the Transcontinental Route. (5)* It was also reported that the road from Phoenix to Yuma was "fine" thereby adding another section to the Transcontinental Highway. Eleven hours and ten minutes was the time by car from Phoenix to Yuma over 206 miles of desert.

The road to Mammoth Wash area was now being improved, for it contained 3½ miles of heart-breaking sand that stopped many a car. A Charles Bigelow was at work there with a force of men building a roadbed of brush, straw and boards. (6)* Los Angeles was making plans to cover every foot of the road between Los Angeles and Yuma. San Diego had better hurry or it would lose the race for the road through the Imperial Valley.

However, Imperial Valley folks and Ed Fletcher were not idle. On January 17, 1912, a "Good Road's Smoker" was held at the Hotel Oregon in El Centro, which drew a large crowd. (7)* A branch of the San Diego-Imperial Good Road's Association was formed and each person attending was charged one dollar to use for promotion purposes. By now the San Diego-Yuma road was already under construction with some 150 men working at the western end of the valley. The supervisor of the Fifth District, Ed Boyd, was telling everybody in the county to get involved in the road building project. (8)*

By February, 1912, the fight over the route for the so called National or Ocean to Ocean Highway from Yuma west was heating up. Los Angeles was still pushing for a desert route by way of the Salton Sea while San Diego contented that it was a much more practical by way of Holtville and El Centro

A State highway commission was appointed by Governor Johnson, while A. B. Fletcher, a highway engineer, was employed to direct the work. (9)* At the same time, Supervisor Ed Boyd submitted to the commission plans for the plank road which were approved as being feasible for the sand dunes.

In May, 1912, the road to Yuma was being pushed along and it was promised that it would be only a short time before the trip could be made in three to four hours. Over 20 miles of the route had been completed and within another week, 38 more miles were to be placed in shape for travel. (10)* By June, 1912, Supervisor Boyd was able to take an auto trip on the new stretch of road that now reached the sand hills.

The year 1913 started off with a meeting in Yuma of the San Diego, El Centro, Holtville and Phoenix Auxiliaries to the Southern Highway Association. (2)* The meeting was headed up by Arizona George W. P. Hunt, who reported that engineers would be in the field within two weeks to locate the bridge across the Colorado River at Yuma. It was determined to have it cross from the Old Penitentiary Hill on the Arizona side to the Army Barracks (Indian Hill) on the California side, a distance of 447 feet. This was the only practical place where a bridge could be built at a reasonable cost and within a short period of time. The Federal Government, California and Arizona each had appropriated $25,000 for the construction of the bridge. (13)* The proposed road from Holtville to Yuma would cross the sand hills and connect with the bridge at Yuma.

Ed Fletcher had attended a Highway Commission meeting in Sacramento and guaranteed to raise the necessary funds to make the plank road a reality. It was estimated that the road's construction would cost about $25,000. (14)* Now that the bridge across the Colorado was a certainty, the sand hills route offered the cheapest solution to the question of which route to take: Los Angeles or San Diego. (15)*

Since San Diego had promised to furnish 13,000 pieces of 2x12 inch Oregon Pine lumber delivered to San Pedro, California, Holtville had to pay freight to Ogilby, California, and lay the timber hammock fashion across the sand hills. (16)* Even Yuma people got into the act when L. W. Alexander, Secretary of the Commercial Club, and County Assessor A. B. Ming, accompanied by Carl McStay, field representative of the Southern California Association of Good Roads and Ed

Boyd, went out to the Holtville cut off on November 13, 1917. (17)* They inspected the work that had been done and heard about the lumber donated by San Diego Chamber of Commerce. The Yuma Commercial Club (later to be known as the Yuma Chamber of Commerce) was trying to get the Southern Pacific Railroad Company to lower its freight rate on the lumber shipment.

Not much is known about the first plank road, which was started February 13, 1915, (18)* and finished in a short time. The reader must however remember that the plank road was just a small part of the over all road building project from San Diego to Yuma. It was built by free labor and maintained by the travelers themselves.

The historical files of the Automobile Club of Southern California reflect that the road was built of three 2x8 inch planks laid side by side to make a 24 inch wide track. There were two such tracks, as far apart as the width of an automobile's wheels and these tracks were fastened on 4x6 inch ties. However, Fletcher in his memoirs write: "The planks were two inches thick, 12 inches wide, and 12 feet long." Photos show that the tracks did in fact contain three boards side by side with the separated tracks being held together by what appears to be 2x4 lumber.

The road was one lane wide and covered a distance of about six miles over the worst part of the dunes. There were turnouts approximately every ¼ of a mile allowing cars to pass each other. Should two cars have met each other in the middle, one would have to back up. (19)*. This probably made for some interesting and stimulating conversation.

Early in 1914, Yuma was made aware of its importance to the Ocean to Ocean Highway link. "When the new bridge is completed across the Colorado River, Yuma will be the gateway to all overland travel to the Pacific slope. When the highway is completed, Yuma will be greatly benefited and is certain to grow by leaps and bounds as a result," said Harry Locke of Phoenix. (20)* Excursions were being made regularly across the sand dunes, in order to promote the ease and safety of travel. Supervisor Ed Boyd was working hard to make the road a reality, while Arizona and California were rushing the completion of the Transcontinental Road bridge across the Colorado River at Yuma (21)*

By April 2, 1914, a booster's meeting practically assured that Boyd's Yuma Road would become a reality. The meeting was held at the Davis Hall in El Centro and two resolutions were adopted: one to ask the County Supervisors to appropriate the sum of $3,500 for paying freight of the lumber for the sand hills and if the supervisors did not see fit to get the money, locals would dig into their pockets to guarantee the amount. (22)*

At the same meeting, Fletcher read a telegram from the state engineer stating that the road from Coyote Wells to Mountain Springs had been planned and surveyed and bids would shortly be put out for construction. On April 7, 1914, the county solon appropriated $500 to defray expenses to survey the Yuma road across the dunes, stating that more money would be available if the route proved feasible. Voting for the funds were S. J. Seat, Newt Gray, and the ever present hard working Ed Boyd. (23)*

By April 23, 1914, $4,587 was raised for the shipment of lumber and Supervisor Boyd telegrammed San Diego that the money was available. (24)* No delay had been expected as Imperial County Surveyor C. L. Perry was satisfied with the survey, but did suggest that the road be coated with oil between Highline Canal and the sand hills, and asked the county for an appropriation of about $1,700 for this work. (25)*. The county supervisors, however, changed their minds in May, stating that the road was too big a deal and too much money for the county treasury. (26)* Supervisor Boyd, disheartened as he left the meeting declared: "It is apparent that the supervisors did not want a road through the valley."

Fletcher also had bad news for the road boosters when he said that Fred C. Jackson had turned against the Yuma Road. He declared: "You can tell the people of Holtville, that I am still working for the Holtville-Yuma road with all my might and that I will continue to work for it until the people of Holtville tell me to quit." (27)* But Ed Boyd had not quit either, as he was still collecting small sums of money here and there for the lumber shipment. Research has not revealed what happened to the $4,587 that Ed Boyd had already collected for the lumber shipment.

It was expected, even with all the problems, that work on the road across the dunes might be completed by November, 1914. However, by September 5, delay in shipping the lumber was made known to the newly appointed superintendent of the Yuma Road, Clayton R. Taylor. San Diego folks stated that it was impossible to ship one third of the lumber as ordered and that the whole load would have to be shipped. (28)*

Supervisor Ed Boyd was still urging the completion of the sand hills road in order to prevent a death trap which existed. "If the road is built, chances of a man getting lost is reduced to a minimum, since automobiles would pass over the road constantly," said Boyd. It seems that a John Bowers and his 14 year old son had undertaken to cross the sand hills and ran out of water, the father dying in the sand before he could be found. The son survived. (29)*

Work on the road was finally started when Newt Gray took a load of water tanks to store water for the men hauling lumber. A well driller was expected from Yuma to drill a new well at the edge of the dunes while teams were hauling lumber from Knob Station on the Southern pacific Railroad to eastern edge of the sand hills. (30)*

Once more Boyd took it upon himself to impress upon the locals the need of immediate action in completing the road. Since Clayton R. Taylor had resigned from his position as Superintendent of Road Construction, matters were at a standstill. (31)* In November, State Engineer C. L. Clark of the Highway Association and his engineer went over the sand hills route and reported back to the State Highway Commission who then had to decide between the sand hills road and Mammoth Wash Road. (32)*

On November 16, 1914, the Imperial County Board of supervisors passed a resolution asking the California State Highway Commission to include the road from El Centro to Yuma in the state's system. The resolution noted that Arizona had voted bonds for a good road east of Yuma. The National Government, and Arizona were building a great bridge across the river. In addition they pointed out that the state's system would not be completed unless the plans included the highway to the state line at Yuma. Supervisor Boyd, John Boyce and J. J. Carr voted for the resolution while Arthur Shepard and W. F. Beal voted against it. The final result was the inclusion of the desert road in the state's road system. The county officials called it the Boyd Road. (33)*

Now that the Holtville-Yuma route was assured, Boyd, returning by rail from Sacramento, brought news that $200,000 had been promised for a concrete highway between El Centro and Yuma by way of the sand hills (34)* although he was not able to say when work would start. He reminded the listeners that it could take at least two years before work actually would begin on the concrete road from San Diego to El Centro. History shows that the concrete road was not built until the 1920's.

Once again Boyd had to urge the valleyites to rush the plank road as the San Diego Exposition would be held in 1915, and the travelers would need a better road. The matter of the lumber came up again when Boyd said

Left to Right: Col. Ed Fletcher,
John M. Eshlemen, Phil D. Swing
and D. C. Bitler, Imperial County
District Attorney, 1915

Building the first plank road across the Algodones Dunes, about 1915.

that there was plenty of money available and wondered why hadn't they started working on it yet, adding that he had money on hand to pay freight for lumber to Ogilby and then to the sand hills. Boyd's figures were: Freight $1,300, wire and bolts $500, hauling lumber to the sand hills $650, hay for the teams $250, and groceries for men $250, which made a total of $2,900. He had $3,500 on hand. (35)*

C. W. Burton, another chairman of the Yuma Road Committee resigned due to the committee's failure to order the planks that were waiting at the Port of San Pedro, California. (36)*. It wasn't until December 11, 1914, that the planks were actually ordered by the new committee, who advised that three miles of planks were to be laid cross-wise in the sand. It was understood that holes would be bored at the end of each board, six inches from the butt to permit bolting or wiring. Since the lumber belonged to San Diego, they would be able to salvage it when the state built a concrete highway. (37)*

The committee had $4,300 on hand and were promised $6,000 from the county as soon as the state reimbursed the county for the amount they put into the Yuma bridge. A check to pay for the planks was issued March 10, 1915 to the McCormack Lumber Company, Edward Fletcher Trustee. (38)* General Manager Ed Boyd awarded the hauling contract to C. L. Gillett and C. G. Frisbie (39)* for the estimated four mile distance. Boyd said it would cost an additional $300 to smooth over the sand before the planks could be laid.

One hundred men from the surrounding Imperial Valley donated a day a week for labor on the road. Beef was offered from Vaughn Azdehrian and F. T. Hagg who also said they would barbecue it free. A big traction engine was donated by No. 7 Water Company and it was used to cross the mesa to find soft spots.

The following ad was placed in the February 8, 1915 **Weekly Holtville Tribune** newspaper by Ed Boyd as general manager of the Yuma Road.

"All together now to task in laying the sand hill planks. Get your hammer in action. The lumber for the Yuma-Imperial Valley Road is back up the sand hill, ready for you to build. Here is the plan of action. Three miles of lumber we are to lay then (sic), is divided into five sections. Each of the five older towns in the valley, Calexico, El Centro, Imperial, Brawley and Holtville are to send a four horse team and six men should build one of the five sections in about 10 days. The road committee has the money raised by private subscription and an appropriation from the county supervisors to furnish you all the wholesome bar-b-que meat and other grub and feed for the stock. There will be no wages paid. The lumber is ready for you and the sooner (after Wednesday) you get at it the sooner the road will be ready to use. Yuma has a big celebration March 15th to formally open the Yuma bridge to travel, and we want to open the road the same day by going to their celebration over the Ocean to Ocean Highway. For further information see Frank Reading, El Centro, A. M. Shenk, Calexico, or anybody in Holtville."

The first spike was driven in the construction of the plank road on February 13, 1915. (41)* A camp was established at Gray's Well, named after Holtville's Newt Gray, however, the camp itself was called Camp Boyd. Lt. Governor J. M. Eshleman of California, Charles Berkhan, Fred Place, J. V. Wachtel, Sam Dunaway, B. Solomon, Albert Dunham, Roy Ramsey, and others took turns at driving the spikes. (42)*

On February 17, 1915, Robert Goree, manager of Edgar Brothers, a Calexico store, requested automobile drivers to pack down the mesa road west of the sand hills. He thought that 50 or so automobiles going over twice would do the job while the road was damp. (42)* No record was found of what sort of traffic jam this must have caused on a one lane dirt road.

February 20, saw the completion of a temporary road by scattering planks in the worst places. Many men were on the job along with Ed Boyd and George H. Johnson, who had taken a number of men in his car. (44)* Already men were thinking of how to cash in on the new road. George H. Johnson of Holtville and Charles Crombie of El Centro were in Yuma on February 26, 1915, stating that they were going to establish a stage line from El Centro via Holtville to Yuma. It took them 5½ hours in a Studebaker Six car for the trip. They found the board road fine, just like running over a concrete road surface. After establishing an office, they stated that they were expecting to make daily trips each way from El Centro to Yuma. (45)* Another group consisting of S. P. Sibly, A. Betham, Arthur Norman, R. R. Hale, and W. W. Sippel, owner of a new White Star Stage route, made a 6 hour 15 minute trip across the dunes to Yuma and reported that the road work was progressing nicely. (46)*

By March 5, it was surprisingly reported by Fletcher that 40,000 more feet of lumber was enroute to the Imperial Valley, and now the whole bad portion of the dunes could be planked. (47)* Without additional work by March 11, the sand hills road could have been as it stood, as about 95 percent of it was safe for the thousands of cars that were expected to pass over it. The route was now boosted by Harry Locke, mapping engineer of the Borderland Route Book, which was soon to be published and placed in all the principal hotels and garages along the Southern National Highway for distribution to tourists. Locke was directing motorists to take the Yuma Road route, the shorter more direct route which eliminated the Mammoth Wash Road, (48)* long a dream of boosters of the Southern National Highway Association.

Driving records were being set crossing the desert from Holtville to Yuma via the sand dunes every day with speeds of 40 to 60 miles per hour reported. Perhaps these speeds were exaggerated as Boyd himself traveled from Holtville to Yuma in two hours and 55 minutes with 20 minutes to cross the dunes, (49)* a more believable time.

To make passing easier on the one land road, turnouts were constructed half a mile apart. (50)* Virgil Sam of Yuma then cut Boyd's driving record to two hours and 20 minutes from Yuma to Imperial in a big Mitchell Racer car. In retrospect, it took him five hours and 37 minutes via the Mammoth Wash route. (51)* Another record was claimed by a motorcyclist, Guy Urquhart, who reported making the trip in two hours from Holtville to Yuma. "A motorcycle can make a 30 mile speed easily on the planks," he said. (52)*

A citizen of Yuma, Arizona, A. J. Eddy, made an auto trip to Holtville and El Centro on March 25, 1915, (53)* with a couple of eastern automobile tourists and had this to say about the plank road. "That part of the plank road that is finished, is in excellent condition. There is no sand on the road to bother. There are about two more carloads of planking to be laid, to finish the road, which so far has proven a great success as an automobile thoroughfare. The people of El Centro and Holtville are looking forward to the great bridge celebration and there will be a large crowd in Yuma during the celebration from the Imperial Valley."

Traveling off the planks.

By March 29, 1915, most of the work was finished, except for putting clay for about a mile on the road approaching the planks and several more miles of planks that had to be nailed down. (54)* On April 1, Ed Boyd was bordering on happiness and half unrest as he had worked himself out of a job. The plank road to Yuma was finished and in shape for all kinds of automobiles to run on. Only about a mile of the road on the eastern end of the sand dunes had yet to be covered with clay and brush as a surface. The $6,000 appropriation by the county for the Colorado Bridge had not been touched, and it was hoped this would take care of the maintenance of the road for the next two years.

All the credit for building the road was given to Boyd. For three years he had worked, almost tirelessly, to get his dream of an eastern outlet for Imperial Valley realized. It was estimated that during that time he had spent $2,000 to $3,000 of his own money. He also was out two automobiles and put in many days that could have been put in private profit for himself. He was a credit to his county and someday the people of the valley will build him a monument to keep the memory of the man who dreamed and fought for his dream of Imperial Valley. (55)*

Between 1908 and 1927, Henry Ford sold 15 million "Tin Lizzies"... The Model T was a popular model as shown in this photograph of Main Street, Yuma, Arizona.

The many friends of Ed Boyd appreciated his work and on April 3, 1915, surprised him with a brand new automobile. The new Ford was bought as a token of his work in connection with the sand hills road to Yuma. Boyd had parked his old car downtown and while he was out of sight, his friends replaced it with a brand new one. A sign on the door read: "This is for you, Ed." The rear of the car was decorated with a big silver flag. The old car was not returned to him but was given to Newt Gray who also had worked untiring with Boyd. (56)* Now that the road was completed, Newt Gray saw an opportunity to be of great help

(and also make money) to the motorist traveler by establishing a small store at the county well, supplying gasoline, automobile parts and repairs as well as groceries. (57)* As it will be later seen, he also furnished them liquid refreshments which would get him in trouble with the law.

Two hours of suspense and anxiety were endured on a March afternoon in 1924 as the structure moved in a true straight course towards the north abutment. The measurements for distance were so carefully made that the anchor bolts set in the abutment fitted the holes in the shoes of the bridge to within one thirty-secondth of an inch.

Surmounting numerous construction problems, the Ocean to Ocean Highway bridge was constructed on the Arizona side of the Colorado River. In due time, the Omaha Steel Works was ready to pull the bridge over the river to its permanent position. The least accident or mistake could cause a catastrophe and hurl the entire structure into the river.

THE COLORADO RIVER WAGON AND AUTO BRIDGE

On April 11, 1915, an excursion was organized consisting of 50 or more cars from Yuma to Imperial Valley. To celebrate this momentous occasion, a new cheer was written. (1)*

The reason for the convening of this happy group is undocumented but they certainly could have been on their way to celebrate the opening of the plank road.

"Y-U-M-A"

Yuma

Where? There.

Arizona

Who are we? Don't you see.

Are we in it? We should smile.

Boosting, Hustling, all the while.

RAH, RAH, RAH
ZIZ BOOM BAH.
And here to stay - don't you forget.

The Yuma People come to town,
Riding in automobiles,
They crossed the bridge,
so could not drown -
With live ones at the steering wheels.

We love you California,
And built the bridge to reach you.
With big brass bands and living hands,
We came across to greet you.

We love you Imperial County,
And thank you for your greeting,
From Holtville to El Centro,
And Brawley to Calexico.
The Eden of the Nation.
We want you all, both the great and small,
To come to our celebration.

The Yuma bridge was nearing completion when on April 18, 1915, a car driven by L. L. Odle of Yuma Indian Agency was the first to cross it. Harry Hill and W. G. Crowder of Yuma were the next two to cross. It was estimated that it would take about two more weeks to complete it for the general use of the public. (2)*

The Yuma bridge committee of the Yuma Commercial Club had completed specifications for the big electric sign that would later be seen for miles. Commercial Club Secretary L. W. Alexander advised that it had to have the word "YUMA" and "OCEAN TO OCEAN HIGHWAY" in letters six feet high, with a proportionate width, and with the letters "YUMA" four feet high. The letters were to be constructed from 24-gauge galvanized sheet metal, equipped with the latest approved sign receptacles and wired with double-braided rubber covered wire. The frame had to be of angle steel, securely bolted and fastened to the bridge in a perfectly secure manner. The sign was 192 feet long with the letters facing California and Arizona. The paint was a beautiful white of two coats of noncorrosive paint and two finishing coats. It would use 650 five watt Tungsten lamps, with weatherproofed sockets or receptacles. The switch to turn the sign on and off was located on the Arizona side of the Colorado River. (3)*

The contract to build the sign was won by N. S. Sparks, a Yuma man. He estimated that the work would cost about $998 and he would finish within 30 days. (4)*

Lighting the letters on the bridge was short-lived. The monthly assessment to pay for the electricity quickly proved unpopular with the business community.

On April 17, E. A. Starbuck of Holtville received a telegram from the American Automobile Association of New York, stating that they had approved the sand hills route.(5)* Then by May 1915, California Assemblyman R.W. Willis reported that the state legislature has passed a bill allowing $100,000 for improvement of the sand hills road and the like amount for the Salton Sea Road.(6)*

Remember our hard working Newton "Newt" Gray of Holtville? While all this bridge building and road opening doin's was going on, he was arrested on May 3, 1915, for violating Holtville's liquor ordinance by having in his possession a gunny sack of 25 quarts of Maeir beer and twelve pints of whiskey. The arrest was made by City Manager E.A. Starbuck, Deputy Marshall W.M. Cornelison, and night watch Votaw. (7)* On May 7th, Gray was found guilty by a jury for possession of alcohol and received 60 days in the county jail and a fine of $250.00. (8)*

Several hundred maps were distributed by Col. Ed on his visit to the valley and Yuma area in May. They had been prepared by the Highway Commission of California and showed mileages from Los Angeles to Yuma. In black lines it pointed out all the state highways completed and under construction. The distance to Yuma from Los Angeles via San Diego, El Centro and the sand hills route was 310 miles, while the distance from El Centro to Yuma was given as 59 miles. (9)*

The big day for Yuma's bridge opening finally arrived on May 27, 1915~ when "Queen Pauline" touched the electric button at 8 o'clock and the great "Ocean-to-Ocean Highway " sign leaped into life and flamed its welcome across the river. A mighty cheer went up from the multitudes witnessing the event.(10)* Many attractions had been offered to the celebrants. Auto races were held in the morning and horse races in the afternoon. Company D of the 12th U.S. Infantry, under the command of Captain C.N. Murphy, performed drills on Main Street. In a parade of cars, E.F. Sanquinetti, a Yuma businessman, won a beautiful loving cup for his Cadillac touring car which was elaborately trimmed in yellow, orange, and green ribbons. He also had a novel electrical display of small lights, appropriately arranged to form the word YUMA. (11)*

In the very late afternoon, visitors from California began to arrive. Hotel Arizona had been designated as the official headquarters for the event with over 500 guests registered. Some of the guests included Lt. Governor John M. Eshleman of California, Governor George W.P. Hunt of Arizona, Col. Ed Fletcher. and a George A. Brook of the Los Angeles Chamber of Commerce.

Prior to the lighting of the sign, the vast audience sang the Star Spangled Banner. Speeches were made by Governor Hunt, Lt. Governor Eshleman, and Col. Ed Fletcher who had been a strong figure in the building of the new bridge. (12)* The queen was the little Pauline Michelson of Yuma, who was most tastefully and richly costumed. She was ably assisted by Yuma ladies Florence Shelby, Majorie Odle, and Frances Ketcherside. (13)*

A sad note to the opening of the bridge was that the "old ferry suicided." It had been the means of crossing the river for the past 40 years and had been tied up near the shore. During the glory and pomp of the new bridge celebration, the old boat quietly sank under the waters of the muddy Colorado, while the vast throng were cheering their new love. It seemed that mortification and shame cast on the old ferry was too much for it to withstand and so sought solace in a suicide's grave. (14)*

Pedistrian and vehicular use of the Madison Avenue Southern Pacific Railroad bridge was strongly discouraged leaving the ferries as the only means of crossing the Colorado River until the Ocean to Ocean Highway Bridge was constructed in 1915.

The inadequacies of the first plank road quickly became apparent. Its successor was constructed at a fabricaiton plant at the Ogilby station. The 8x12 foot, 1,500 pound sections were hauled from Ogilby to the road site - 1916.

Unloading the sections for the second plank road.

The sections were placed butt to toe and bolted down with iron straps.

Time and progress march on but horses were still being used to construct the new road.

IMPROVING THE SAND HILLS ROAD

Turn-out on the second plank road.

Now that the road was completed, folks were using the road across the dunes quite often and regularly. Col. Ed Fletcher even tested a Dodge car covering the trip from Yuma to El Centro in three and a half hours. The driver was a man named Nelson who drove the stock touring car as a pro.(1)*

All types of vehicles were passing over the plank road, such as a vehicle driven by Mr. and Mrs. J.R. Thomas of Jacksonville, Florida. It was a big Hudson Touring car arranged with bedding, camping utensils, and reserving the interior for a house, as they called it.(2)* Was it an early day motor home?

Since the road had been recognized by the State of California, five surveyors were sent out to survey a new road through the sand hills from Holtville to Yuma.(3)* Indications were that the state highway would go straight east from Holtville to Yuma, leaving out the Bond Corner route, which had been agreed upon a year earlier.

Engineer A.N. George began surveying the route as controversy about the Bond Corner continued. The plan of the commission was to remove as far as possible all the terrors of the sand hills before the beginning of summer heat. It was determined that the new road across the sand hills would not follow the first road. (4)* J.G. Beekler of **The Calexico**

Chronicle suggested that a stage line be planned to run every day to keep the road in shape and also assist in advertising the fact that there was a highway via Bond's Corner. (5)*

"Bids were being opened for the lumber with which the sand hills road was to be planked,!' reported the California Highway Commission on February 9, 1916. By February 18, a bid of $25,000 from R.K. Kood Company of Los Angeles was accepted. (6)* Engineer W.W. Patch stated that a total of 1.2 million feet of lumber would be used and six miles of plank road would be built. The road was to be eight feet wide and each individual section 16 feet long. Turnouts of 16 feet in width were to be installed every 1,000 feet.

By March 16, 1916, the new plank road across the dunes was being built. Planks measuring 4 x 6 inches and 4 x 8 inches plus a large number of bolts had arrived in Ogilby, the hub of building activity. The road's completion was of great importance because in March of 1916, a party of people nearly perished in the desert due to lack of water. After the vehicle stalled, all of them, except Maltede, started back to Holtville on foot. They were overcome by thirst and fell by the wayside. A young El Centro couple found them about a mile from the Highline Ditch and took them to the Highline Store. The story of the people's plight was passed on to the sheriff who found Maltede almost unconscious. He had been digging in the sand for water with his bare hands.(8)* 29

Supervisor F. Shaw of Imperial County, after having made a trip over the dunes in May, had this to say about the old plank road: "I would not advise anyone to make the trip, although the road is fairly good and we came in from the Well and Highline Ditch in 55 minutes." He further said that laborers were laying 600 feet a day on the new planks and that it would be completed in a month at most. He explained that the road was on 4x 6 stringers and on this, 4x8 inch planks were bolted through 1 1/4 inch wide strap iron. The whole section was as solid as one piece of timber. The sides of the road were crowned and oiled along the floor and made as solid as possible. About two miles had already been completed at the time he had made the inspection trip. The old planks were pulled up and the county used them to lay over all damp and soft places, this according to Supervisor Shaw's report.

On June 15, 1916, State Engineer Patch reported to Shaw that except for the oiling, the new plank road was completed. The edge of the new road was to be oiled with 65 percent oil and the planks with 95 percent oil. (10)* Work, however, was stopped temporarily due to the extreme heat, but it was expected to resume in the fall. About 3 1/2 miles of the worst places had been planked and about 20 miles of the road in the Imperiai Valley had been oiled. (11)*

Then M.Y. Emmanuel, who had made an inspection of the valley roads in July, said that five miles of the plank road was completed and in excellent shape with the exception of a short piece on the eastern side. it appears that the road crew did not wait for the weather to cool down, but continued building the road. According to Emmanuel, the entire surface was coated with heavy oil and gravel, making the road practically free from vibrating ripples. The road according to him, constructed of 4 x 10 inch and 4 x 12 inch timber laid on three based supports of girders. These surface timbers were laid horizontally and strapped to the girders by three iron straps in section of about 12 feet long. The roadbed was eight feet wide with 16 feet turnouts approximately .2 miles apart. (12)* The road to Yuma from the east end of the new plank road was fairly good although through the Indian Reservation the road was badly cut up and dusty and broken springs could be expected on heavily laden cars over this stretch. (13)*

Newt Gray, however, disputed this bad story, explaining that any desert road was bound to be a little sandy, but he maintained the road could be traveled up to 10 miles per hour. He suggested that one should do as he did: take along enough provisions for all emergencies, including a big sand belt, in case a car should strike the soft sand. He also took along a 100 pound block of ice to cool the well water and make the desert pretty passable after all. (14)*

A well had been dug about 30 miles east of Holtville and about 20 miles from the Highline Canal. The plank road which was pictured in most of the National Scientific magazines ran within 200 yards of the well. It was one of the bucket - drawing type and furnished cool water. To the traveler stopping at this desert oasis, it was a rare treat indeed.(15)*

The winter rains came at the right time to produce stands of dune primroses.

Time to practice your tricks while waiting.

TRAVEL ON THE OLD PLANK ROAD

Today a modern, divided, multilane highway allows lines of passing motorists through the ever shifting Algodones Dunes. Except for the usual scenery of sand, utility poles and some vegetation, the highway looks just like any other freeway anywhere.

But not so long ago, traveling across the sand dunes was sometimes an adventure. The road inspired its share of tall tales. One such story was about a motorist, who after a dust storm, spotted a hat laying on the sand off the plank roadway. He stopped to pick up the hat and was amazed to find a man's head sticking above the sand. "I'll get a shovel and dig you out." said the Good Samaritan. "Better go back to town and hire a tractor," replied the buried man, I'm sittin' at the wheel of my Model T. (1)*

Even with the new paved road, nature occasionally reminds travelers of its power. Strong winds at time lash Interstate 8 with sand that can strip the paint off a vehicle in no time at all. Traffic slows down and plow trucks have to clean off the shifting sand.

One of the worst desert storms in California was encountered by the motorists in the sand hills on March 21, 1924. The storm, which had begun in the afternoon, increased in velocity by midnight, forcing the autoists to stop and brace their cars. Those who were within

On the back of this snapshot is the inscription: "Sand storm in sand hills when Dad, Jay and Bunk stayed out there all night. 1924

close proximity of Gray's Well considered themselves lucky. There the only shelter was a water tower and a few small buildings. But it proved a wonderful relief against the wind in keeping the elements from literally tearing the covers from the sides and tops of their machines.

"Under the light of the bright moon the wind tore the fine sand of glittering glory that obscured even the asphalt under the beaming headlights of the motors. The sand stung and lashed the cheeks of individuals, buried automobiles from sight and forced their owners to seek refuge from the torment," wrote **The Morning Sun** newspaper of Yuma, Arizona. (2)*

Sand drifts ten and fifteen feet high were piled on the plank road and in the open spaces between the hills and the road, throwing up a barrier that proved absolutely impassable.

The pavement on the other side of the planks was cleared easily, but there were places on the plank road where sand was retained by the tons. Traffic needless to say was severely curtailed. The night spent on the desert by the stalled travelers was declared to have been one of the most miserable ever experienced by them. A warning was posted on both ends of the dunes that no one should cross the stretch of road until the storm had completely spent itself.

Dan Woods, a reporter for the **Tri-Valley Dispatch** newspaper, wrote of the travel on the plank road as being so rough that it would give the motorist a terrible headache. If one would get off the planks, there was a lot of back breaking work to get back on. While Woods' family was traveling on the planks one fine day, the kids got a nose bleed from the bouncing and jolting. The only remedy his father knew was to hold a dime in the roof of one's mouth. But Woods' father said he couldn't afford to lose the money by the kids swallowing it, so they put rocks in their mouths which seemed to work. (3)*

Anyone who ever crossed the dunes in an automobile when there was a breeze of any magnitude will remember the eerie feeling produced by the sight of the drifting, swirling sands. It appears as though one were traveling crosswise in a running stream of water. Just imagine how it was in the good ol' days, especially in cars that had no side windows to roll up and were of canvas top.

A journey over the dunes was sometimes hazardous because the road was vulnerable to shifting sand. During some sudden storms, sands drifting across the road completely blotted it out. If the motorist carried a shovel, as some folks did, he would shovel off the sand and continue on. If not, he was in trouble with no way across, because upkeep of the road was sporadic and expensive.(4)*

Some say today that one might find old freight wagons and autos buried in the sand by the dunes continual shifting.

Water also posed a problem in the sand dunes. Since the nearest water was only available in the distant Imperial and Yuma Valleys, it was first planned to pipe this water to the plank road. Before the final plans were made, an attempt was made to find water in an open valley in the dunes. Drilling started and to the amazement of those who scoffed at the idea, a flow of 500 gallons was struck at 153 feet, thus removing the water hazard.(5)*

Ed Lehnes, a reporter for **The Yuma Daily Sun** newspaper, wrote that the second plank road was rugged and rough. He said that at 20 miles per hour, the chances of bouncing off were very good and that drivers made as many as 30 stops in order to prevent damage to the vehicles from violent shaking.

Rough indeed was the trip across the dunes on the first plank road. A trip out of Holtville to Yuma quickly turned from a road to a narrow sand track where the creosote bushes had been chopped out to make a roadway. When the sand got softer, the driver of a car would have to stop and deflate the tires, which in some instances would give it better traction. Many motorists carried strips of tarp or just cut down bushes, placing these under the wheels if they happened to have gotten stuck. Since engine boil-over was always a problem, local motorists often carried water and spare parts, especially fan belts. There were instances when one's pants belt doubled in a pinch.

As the motorist neared the sand hills, they must have appeared unsurpassable to him. To climb the hills, a driver would skirt the dunes on the windward side, following the crust, as the sand seemed to be packed harder here then on the leeward side. Now he did not dare to stop, for if he bogged down, there wasn't even a bush to cut down to put under the wheels.

Eventually the weary traveler would reach the plank road that was under construction. Even then the wind was beginning to drift sand over the road. Once the sand hills were crossed, it was now only ten more miles to Yuma.

A warning had been given to motorists going through the sand dunes in 1919 by the El Centro branch of the Automobile Club of Southern California. The club's spokesman stated that the planks were dangerous and several cars had been badly damaged. "Cars are injured in the drive, engines are cracked and shattered, and in many cases the machines have to be pulled many miles by a team of horses." concluded the spokesman.(6)* The club also warned of death and danger to people. Travel had been light over the road and motorists who tried to cross, often suffered from thirst and sometimes were in danger of death. There was little chance of succor arriving unless a call for aid reached either Holtville or Yuma.

The plank road was one of the funniest and most exasperating roads imaginable.(7)* It went up and down like a toboggan and rattled all those Model T Fords. Sometimes on the crest of a dune, a driver was able to glimpse another car approaching far ahead. That would start a race for the nearest off shoot siding known locally as "God-Damns" and they occasioned as many blows as curses when two cars met between turnouts, as one would have to back up, a chore even on a good road.

By September, 1916, travelers had already complained that the plank road was buried under sand and practically useless. San Diego and Imperial County officials stated that the lack of funds precluded keeping crews on the job after the California Road Commission withdrew its men and teams. (8)*

It had been estimated that the road would have to be moved every three years. The sections could be rejoined and placed where ever desired. Small amounts of sand were removed by horse pulled scrapers, but when a large dune went across the road, the most economical way was just to move the road.(37)*

Los Angeles officials weren't helping matters either when they stated that the new road was half the time submerged and impassable.(9)* This however was not true. It was the old plank road that was often completely covered by blowing sand that would pile up in the center portion between the two board tracks. The new road was solid and the sand just went over it, well most of the time anyway.

To make travel across the dunes enjoyable by 'leaving the driving to others", stage lines were periodically inaugurated. One such was the auto line H.O. Walker and Charles Johnson of Holtville. They purchased a seven passenger Reo and since Johnson was an old hand on the planks, the trips were quite comfortable. Johnson probably was the first man who had a stage across the first plank road. (10)* He and Walker had a through line from San Diego to Campo and there transfers were made to a stage for a direct trip to Yuma. From El Centro the stage left from the Barbara Worth Hotel.

With the improvements of the car, it was inevitable that they would be used commercially. The word "jitney" came into use, meaning individually owned and operated cars, which carried passengers along regularly established routes. Expensive cars such as Cadillacs, Locomobile, Packards, Pierce-Arrow, Stevens-Duryea, and Wintons were usually selected and modified for additional seating. Stage service to the Imperial Valley began as early as 1911, when Charles W. Griese of San Diego went into business. By 1924, there were daily services: $15.00 for the 9 1/2 hour trip from San Diego to El Centro.

The Imperial Valley Stage Line advertised that their equipment was the "best money could buy. with the maximum of comfort and convenience, which is offered to every passenger who takes the trip of unparalleled scenic beauty."

The stages were hardly roomy and the passengers might have to sit on a fender or running board yet day after day and often into the night the stages ran. However the jitneys were abruptly outlawed, after the California

Railway Commission assumed the regulations of bus companies in May, 1917. The Imperial Valley Auto Stage, because it left from the front of the San Diego Pickwick Theater, became known as the Pickwick Stages and by the end of 1925 evolved into the Greyhound Bus Lines. (11)*

One man who rode the stages had this to say about it: "The trip over the desert is really enjoyable, commented Sim Freund, a prominent Yuma merchant. "It is really remarkable and no discomforts were felt and from the time one leaves El Centro and arrives in Yuma, he is impressed with the trip more and more. The new plank road is really remarkable, carrying autos over what are known as sand dunes in the desert. It is no ride across the Sahara Desert, but rather a pleasure trip any person can enjoy." (12)*

Another stage line was formed, this one by two Yuma men, Dale Zumwalt, a druggist . and Oscar Ling. (13)* The stage left from the Arizona Hotel in Yuma and made two trips a day. A large Pop-Hartford car was used and made the trip from El Centro to Yuma enjoyable and as fast as conditions would allow.

H.F. Morgan, who had a service station in the sand dunes, reported in May 1917, that the planks were clean and said that the road was not only passable, but in most excellent condition. Workmen had labored hard to remove the sand that in places was very deep. After the men were recalled, the sand once again started to build up over the road. (14)*

One man from Dallas. Texas, J.P. Smith, had harsh words for the sand hills route in 1917.(15)* He had rescued a man and his son from death, which he reported when he arrived in Yuma looking like a faded dishrag. But his jocular disposition made an interesting graphic description of the trip over what he called "the sand dunes from hell."

"Why, someone who must have either wanted to do us harm or desired to see us perish outright, directed us over the plank road from Holtville to Yuma. This is well nigh impassable between here and Holtville. We ascertained that had we come on the Niland Road, we would have had little or no trouble. As it was, I counted 17 deserted automobiles all the way from a Cadillac Eight to Ford Roadsters between the plank road and Holtville. What became of the occupants of this string of automobiles, only God and the burning sand can reveal." It sounds as if he really was not complaining about the planks themselves, but only the road between Holtville and the western end of the plank road.

After 1919, the trip from Holtville to Yuma or the other way. usually took a long time due to the increase of traffic. It was reported that many a fight broke out over who had the right of way, which in turn caused considerable traffic jams. Often it took many hours to back up the jammed cars to a turn-out.

Reminiscing about the travel over the plank road in 1925, Diana F. Hudnal wrote a brief account of it in the **San Diego Tribune** newspaper. (16)*

“After checking gas and filling up with water, off we went into low gear. You chug, jig, jog, and feel every plank. Every bone in your body works loose and with us sitting on the hard floor, we ached all over before the thing ends. Now and then you pull over to a single siding to let a car coming from the other direction pass, then on you go and keep on going to reach the next siding before another car comes your way.”

“All along you notice nuts, bolts, screws, fenders, headlights, tailpipes, mufflers, and anything else that can work loose and fall off. But no one dared to stop and pick them up because of the treacherous bottomless sands.”

“Completely exhausted and literally blinded by the sand glare, we missed the side stop and met head on with a group of five cars. The drivers all stopped and came forward to survey the mess of messes blocking the highway. Silently they looked in disbelief at the dirt covered individuals sitting unprotected from the glaring sun in a tiny car of questionable type with the rags dragging along the bottom. After a brief huddle, what to do with us, these business men dressed in fine clothing, decided to chance the shifting sand long enough to get us out of the way. Lifting us in the car bodily, they set us off in the sand, went back to their cars for a short distance, stopped, came back and while we sat there wishing we could commit some sort of suicide that was not messy, they put us back onto the miserable section of the National Highway and we kept going. At the other end of the road was a large garage with a large sign: “General Overhaul.”

Gray's Well on the west side of the dunes. Not all the cars that challenged the dunes were successful. Perhaps this photograph was taken the same spring as the one on page 40. There is a good showing of Ajo lilies. ***Hesperocallis undulata*** *and dune primroses,* ***Oenothera avita.***

PLANK ROAD
TRAFFIC SIGNS AND REGULATIONS

Road Control

It is impossible to find anyone who remembers if there were any road signs or controls over the plank road portion of the Ocean to Ocean Highway that was also known as the Old Spanish Trail, Transcontinental Highway, Border Route Highway 80. Now of course it is known as Interstate 8.

The first reference of any kind of control on the plank road was during the road opening celebration in 1915.(1)* the controls stated:

10:30 to 11:30 westbound
11:30 to 1:30 eastbound
1:30 to 4:00 will alternate 45 minutes each way
4:00 to 7:00 westbound
7:30 to 10:30 eastbound

Traffic was allowed to only go in the direction stipulated. Officers were present to prevent any one trying to go the wrong way. It was suggested that it would be the motorists fault if they had to wait hours to get through. One very prominent sign was an upright pole with a tire mounted on it announcing to the motorists that a turn-out was near. This particular sign did not appear until the second plank road went into operation.

Not much is known about any other regulations on the first plank road, but by 1916, on the new solid planks, new innovations were being perfected. One was a sign that Newton Gray of Holtville had put out to prevent drivers from becoming lost. Once a week after the frequent wind storms,(2)* Gray would drive his heavy wagon over the road in order to show the track.

The first mention of law enforcement on the plank road was made May 19, 1917.(3)* It all started when Mr. & Mrs. Hogan, traveling from Yuma to the coast, encountered some problems. About half way between Yuma and El Centro, one of their wheel bearings burned out. Having no spare parts to repair it, they caught the daily stage passing by. When they returned the next morning, they found that someone had stripped the car of everything removable: the casings, tubes and tools. According to the Imperial County Officers, this was not the first time an abandoned car had been stripped. Mr. Hogan felt that the Automobile Club of Southern California should do something to apprehend the guilty parties of this sort of crime. No record was found by this researcher if any officers were placed on the plank road at that time.

Photos of the second plank road show that signs had been placed at Gray's Well and New County Well on the eastern end of the road which read: "Caution. This is a one-way road with passing sidings. If another car is between you and the next sign, pull out and wait. Save Delay and Danger. Use Caution."

The first attempt to regulate the traffic flow on the plank road was made in January, 1925.(4)* It was announced that traffic in the sand hills between Yuma and Gray's Well would be regulated by time intervals. Fences were built at both ends of the sand dunes, and at intervals of 40 minutes, traffic would be allowed to proceed one way and the other in order to prevent any traffic jams on the road that caused more profanity than anything else. The Southern California Highway Patrol summed it up by saying: "All the jams were caused by pure foolishness on the part of the auto drivers who could not use the brains God gave them." Trials supposedly were held for a week by the California authorities, however, since there were no promised signs erected at the sand dunes, the proposed 40 minute interval traffic plan probably did not work out or possibly did not even materialize.

On February 26, 1925, the plank road was blocked for a period of time for a party of motorists coming to Yuma from San Diego to celebrate the opening of a stretch of pavement on the west side of the dunes.(5)* Captain Art Leonard of the traffic squad from El Centro and District Attorney Ernest R. Utley had been in Yuma making plans with the city and county motor squads for the plank road control. This lasted only one day and the following morning things were back to "normal."

"When the traffic over the sand dunes became so crowded with the motoring public in 1924-25, studies were made to determine the feasibility of a pavement or the best type of movable road. Several sections of two way plank road were built, but cost of materials and moving them after a sand storm, was cost prohibitive. These sections then might have been used for turnouts.

Photos of these test sections show them to be well constructed and it appears that large bolts, instead of metal bands across the beams, were used to hold the road section together

By mid 1925, the second plank road was rapidly deteriorating. The California Highway

Commission, after a formal hearing, adopted a resolution fixing the speed limit at ten miles per hour(6)* with weight limit set at 8,000 pounds. Signs were erected warning the motoring public of the new regulations.
The following story is copied here as it appeared in **The Morning Sun** newspaper on March 13, 1926. The reader can decide for himself how confusing it must have been to the motorists.

"Motorcar traffic on the plank road to be regulated by controlled plan: opening west end on even hours. Announcement has been made that traffic control system will be installed beginning Sunday (March 14, 1926) on the plank road over the Imperial County sand hills, whereby the traffic will move in one direction for a specific length of time and then will be stopped for that direction while the flow travels in the other direction."

"Captain Oswalt of the Imperial County traffic squad stated that the plan was adopted at a conference of traffic officers and inspectors in El Centro. The controls will be separated from 9 o'clock in the morning until 9 at night, but the present arrangement is only tentative and may be extended for a time. The plan is briefly this: east-bound will be permitted to go ahead from 9:15. from 11 to 11:15 and so on beginning with each odd numbered hour and for 15 minutes thereafter until 9 o'clock at night. Traffic west bound will be permitted to enter the road at 10:15, at 12 noon until 12:15 and so on beginning with each even number until 8 o'clock at night."

"Traffic will be allowed to enter the plank road at Gray's Well on the west end at 9 o'clock and for 15 minutes thereafter. It is presumed, although not announced, that sometimes previous to 8:30 o'clock, traffic arriving at the east end, bound westward, will have to be held up to permit the flow Arizona wards, west. The last car entering on the west end at 9:15 will be given a traffic flag to deliver to the officer stationed at the eastern end. This is presumed that the officer on the west end will know which is the last car."

"If only one car is there to travel east at 9 o'clock, it may be confusing to the officer at the west end waiting for the flag. In the meantime, traffic will then start over the road from the east end. However, as the last car from the west had been allowed to enter the road at 9:15 and the last one from the east at 10:15 will not arrive at the west end until 11, it will be seen that a car traveling east and arriving at Grey's Well at 9:15 1/2: is due for an hour and 1/4 minutes wait before it can start." Now that the reader completely understands the sequence of events, would he go out and try it?

Second Plank Road Algodones Dunes.

Hopefully this is a friendly conversation
among old friends
in a chance meeting
on the narrow plank road.

A careful study of the detour sign indicates an ingenious application of double duty usage materials.

A monument designating this section of the second plank road as a historic site was erected by the California State Pioneers Society, October 16, 1971, on the west end of the sand dunes.

THE FINAL EFFORT

In 1925, a number of problems began to appear when the traffic increased to about 30 vehicles per day.(1)* Cars ran off the road and got stuck. The short turnouts were insufficient for two-way traffic on the one-way road. Traffic jams were frequent and often long when the right of way was disputed. The road itself was falling apart after 10 years of hard usage and exposure to the elements.

It was noticed that the coating of tar, although it filled all the crevices and made the smooth surface desired, did not prevent the sand from piling up and blocking the road. So the new road, which was too heavy to reroute, had to be cleared periodically by men with horse teams and road scrapers. It was estimated that $35,000 was spent annually to keep the seven and a half mile stretch of plank road passable.(2)*

Studies were then begun to determine the feasibility of a permanent pavement or the best type of movable pavement. After long consideration of the cost and other factors involved, a study of the dunes was made in order to determine the best location and type of road feasible in the sand hills.

A civil engineer, E.Q. Sullivan, was called upon to see what he could do toward providing a practical road. (3)* He spent several months studying the movements of the sand dunes as influenced by the wind at all seasons of the year by driving stakes in the sand dunes, he recorded their movements, finding that the dunes seldom exceeded a certain height and that only the smaller ones were continually shifting. During a severe wind storm, those less than ten feet high would travel one hundred feet: dunes over thirty feet in height did not move more than one inch.(4)* From these facts, Sullivan devised a plan of flattening the taller dunes to a height of thirty-five feet, using them as piers and filling the low areas with sand and capping them with blacktop road. Before any pavement was laid, the newly constructed embankment was thoroughly drenched with water to a depth of three feet, which packed the soft blown sand into a comparatively firm sub-grade. The water was obtained from a well dug in the center of the dunes.(5)*

A contract was awarded in December, 1925, to Schmidt and Hitchcock for the fill and the asphaltic concrete paving. The paving itself was sub-contracted to George M. Souter of Los Angeles.(6)* To prevent wind erosion, the contract also specified the oiling of the embankment to form an oil cake with the state assuming liability for sand blown away during a storm. The price for the asphaltic concrete mix was $7.60 per ton and 14 1/2 cents per cubic yard of sand for building the embankment. When completed, the entire seven mile stretch cost approximately $340,000. The pavement was six inches of asphaltic concrete laid in two courses, a four inch base and two inch surface course. It was 20 feet wide, affording ample passing room for motorist.(7)*

The work was carried on under considerable difficulty and with much discomfort from the terrific heat of the desert. It took about 100 days to lay the asphalt with shifts of workmen often toiling 24 hours a day. However, the greatest part of the work was accomplished during the coolness of the night.(8)*

The Roadrunner, a publication of the Standard Oil Company of California, wrote in November of 1926, that the asphaltic concrete was the best solution for the sand sub-grade. The extremely hot days under the boiling sun, with temperatures falling sometimes 40 degrees in a short time, were the most severe test. Immediately to the west of the sand hills, extending from Gray's Well to the East Highline Canal, was a 22 mile stretch of asphaltic concrete pavement five inches thick. It had been laid in 1922. "This pavement is in excellent condition, and affords proof of the durability of asphaltic concrete under heat and subgrade conditions found in the desert," concluded The Roadrunner.

Putting down the asphalt on the new road. 1926

The opening of the finished pavement on August 11, 1926, marked the completion of the most spectacular paving project in America. (9)* Now the days of pioneering a road across the sand dunes are behind the motorist and he has an easy crossing. Only the sight of the old discarded planking, half buried at the side of the highway, bears mute testimony of the early struggle of men to conquer and make these sands his servant.

Photo Credits

Abbreviations: YCHS - Yuma County Historical Society collection at Arizona Historical Society, Yuma, Arizona
AHS - Arizona Historical Society, Tucson
YCCC - Yuma County Chamber of Commerce
PRC - Patricia R. Callahan
Cover Courtesy of Jack Chavez - Yuma

page	
iii	PRC
iv	PRC
v	PRC
vii	PRC
viii	PRC
ix	PRC
xi	YCHS
xii	PRC
xiii	PRC
xv	map courtesy of the late Don Buffkin
1.	YCHS
2.	YCHS
3.	YCHS
5.	PRC
6.	PRC
7.	YCHS
9.	Author
10.	YCHS
11.	YCHS
12.	YCHS
13.	YCHS
19.	YCHS
20.	top left author
21.	lower right AHS
23.	YCHS
24.	YCCC
25.	YCCC
26.	YCHS
27.	YCHS
28.	top left YCHS
	lower right
29.	YCHS
31.	YCHS
33.	YCHS
34.	top left - Charles Kendall Collection
	YCHS
	lower right - Charles Kendall Collection
	YCHS
35.	top left - Charles Kendall Collection
	YRCH
	lower right - AHS
36.	YCHS
37.	YCHS
39.	YCCC
40.	top left - YCHS
	lower right - YCHS
41.	PRC
42.	YCHS
47.	YCHS
48.	AHS
49.	YCHS
51.	AHS
53.	YCHS
54.	YCHS
55.	YCHS
56.	YCHS
57.	PRC
58.	PRC
59.	YCHS
60.	YCHS
61.	YCHS
62.	YCHS

The Algodones Dunes

1. **The North American Deserts** by Stanford, CA, Stanford University Press, 1957
2. Ibid. Blake names it in 1854
3. Algodones Dunes of Southern California by Robert M. and Kenneth S. Norris Geological Society of American Bulletin April, 1961
4. Ibid.
5. Ibid. (Johns, 1954, pg. 17)
6. Ibid.
7. Ibid. The name of Lake Cahuilla (Blake 1907) is issued to conform with U. S. Geological Survey Usage, but Lake LeConte of Gailey (1902 pg. 12) has clear priority.
8. Ibid.
9. Ibid. pg. 617
10. Ibid.
11. Norris states that nearly all of the Algodones Dunes sand is light reddish brown: intensity of color increases slightly from the north to south, pg. 611.
12. **The North American Deserts** by Edmund C. Jaeger,
13. **The California Deserts** by Edmund C. Jaeger Stanford, CA, Stanford University Press, 1965,

Early Travel on the Dunes

1. **Anza Conquers the Desert** by Richard F. Pourade, San Diego, CA. Union-Tribune Publishing Co., 1971
2. Sand Storms in the Dunes were an early Day Hazard by Roscoe G. Wilson, **Arizona Days and Ways** Magazine, June 19, 1960
3. From Dusty Trails to Super Highways by James D. Blick, Brand Book Five, San Diego Corral of Westerners, San Diego, CA, 1978
   The Butterfield Overland Mail by Waterman L. Ormsby, Edited by Lyle H. Wright and Josephine B. Byum, San Marino, CA, Huntington Library, 1962
4. From Dusty Trails to Super Highway by James D. Blick
5. Ibid.
6. Ibid.
7. Ibid.
8. Ibid.
9. **Gold in the Sun** by Richard F. Pourade San Diego, CA, Union-Tribune Publishing Co.
10. Ibid.
11. Ibid.
12. Ibid.
13. **Yuma Examiner** newspaper, October 9, 1911
14. Ibid. October 18, 1911

Politics and Road Building

1. From Dusty Trails to Super Highways by James D. Blick
2. Ibid.
3. When the Sands of the Desert Grew Gold by Gordon Stuart Pacific Palisades, CA by author in 1961
4. **The First Thirty Years** by Otis B. Tout, San Diego, CA, by O. B. Tout, 1931
5. **The Arizona Sentinel**, May 23, 1912, pg. 1
6. Ibid.
   Written in **The Arizona Sentinel**, December 12, 1912, pg. 1, the headlines read: "Ocean to Ocean Highway to built via Mammoth Wash." The year was 1912 and the Los Angeles-Colton-Yuma Auxiliary for Better Roads was really trying to beat San Diego Chamber of Commerce for the desert road to Yuma. The Auxiliary was planning to raise $100,000 with most of it to be used on the Mammoth Wash section. It had been established that it would cost about $6,000 per mile for the 3 miles of bad stretch through Mammoth Wash. Any monies left over were going to be spent on the section between Mammoth Wash and Yuma. Mammoth Wash is a stretch of sand immediately north of Algodones Dunes. Plans were afoot that Charles Begelow would lay a branch of tule foundation and then surface it with clay or gravel.
7. **Weekly Holtville Tribune** newspaper Holtville, California. The newspapers are now in the possession of the Imperial Valley Pioneers Association at the Imperial County Fairgrounds in Imperial, California
8. Ibid. February 8, 1912
9. Ibid. February 29, 1912
10. Ibid. May 30, 1912
11. **Arizona Sentinel**, June 176, 1912
12. **Weekly Holtville Tribune**, January 9, 1913, pg. 2
13. **The First Thirty Years** by Otis B. Tout
14. **Memoirs of Ed Fletcher** by Col. Ed Fletcher, San Diego Pioneer Printers, San Diego, CA, 1952
15. **Weekly Holtville Tribune**, November 13, 1914, pg. 4
16. Ibid.
    The First Thirty Years by Otis B. Tout
17. **Yuma Examiner** newspaper, November 13, 1913
18. **The First Thirty Years** by Otis B. Tout, The Plank Road by James B. Bates. James B. Bates said that it took 3 weeks to build the first plank road, but this author has not been able to find any record showing that the road building took only 3 weeks. Even Otis B. Tout in his **First Thirty Years** fails to mention anything about a plank road building across the dunes in 1912. The author Bates possibly might have meant the road building in the Mammoth Wash area that was being pushed by the Los Angeles Concern. See footnote #6

19. The Valley Imperial - First Annual Historical Volume by Elizabeth Harris. Imperial Valley Pioneers, January, 1956.
    Letter from Automobile Club of Southern California in San Diego Historical Society files, dated July 28, 1959.
20. **Weekly Holtville Tribune**, February 19, 1914
21. **Yuma Examiner**, February 5, 1914
22. **Weekly Holtville Tribune**, April 2, 1914
23. Ibid. April 9, 1914
24. Ibid. April 23, 1914
25. Ibid.
26. Ibid. May 21, 1914
27. Ibid. May 28, 1914
28. Ibid. September 15, 1914
29. **The First Thirty Years** by Otis B. Tout
    Weekly Holtville Tribune, September 17, 1914
30. Ibid. October 1,1914
31. Ibid. October 30, 1914
32. Ibid. November 12, 1914
33. **The First Thirty Years** by Otis B. Tout
34. Ibid. November 23, 1914
35. Ibid. December 10, 1914, pg. 1
36. Ibid.
37. Ibid. December 11, 1914
    Memoirs of Ed Fletcher by Col. Ed Fletcher
    From Dusty Trails to Super Highways
    by James D. Blick
38. **The First Thirty Years** by Otis B. Tout
    The Yuma Sun newspaper, February 5, 1915
39. **Weekly Holtville Tribune**, December 11, 1914
    Memoirs of Ed Fletcher by Co. Ed Fletcher
    From Dusty Trails to Super Highway
    by James D. Blick
40. **Weekly Holtville Tribune**, January 20, 1915
41. **The First Thirty Years** by Otis B. Tout
42. Ibid.
43. **Weekly Holtville Tribune**, February 17, 1915 pg. 1
44. Ibid. February 20, 1915
    Yuma Sun, February 26, 1915
45. **Yuma Sun**, February 26, 1915
    Arizona Sentinel, March 11, 1915
    Weekly Holtville Tribune, February 24, 1915
46. **Yuma Sun**, February 26, 1915
47. **Weekly Holtville Tribune**, March 5, 1915
48. Ibid. March 11, 1915
49. Ibid. March 12, 1915
50. Ibid.
51. Ibid. March 15, 1915
52. Ibid. March 24. 1915
53. **Yuma Sun**, March 26, 1915
54. **Weekly Holtville Tribune**, March 29, 1915, pg. 1
55. Ibid. April 1, 1915
56. Ibid. April 5, 1915
57. Ibid. April 7, 1915

The Colorado River Wagon and Auto Bridge

1. **Weekly Holtville Tribune**, April 10, 1915
   Yuma Sun, April 16, 1915
2. **Yuma Sun**, April 9, 1915
3. **The Arizona Sentinel**, April 9, 1915
   Yuma Sun, April 16, 1915
4. **Yuma Sun**, April 16, 1915
5. **Weekly Holtville Tribune**, April 17, 1915
6. **The First Thirty Years** by Otis B. Tout
7. **Weekly Holtville Tribune**, May 3, 1915
8. Ibid. May 13, 1915
9. Ibid. May 26, 1915
10. **Morning Sun** newspaper, Yuma, May 28, 1915
11. Ibid.
12. Ibid. May 28, 1915
13. Ibid.
14. Ibid.

Improving the Sand Hills Road

1. **Weekly Holtville Tribune**, June 8, 1915
   Yuma, Sun, May 28, 1915
2. **Weekly Holtville Tribune**, October 4, 1915, pg. 1
3. Ibid. December 31, 1915, pg. 1
4. Ibid. January 4, 1916
5. Ibid. January 25, 1916
6. Ibid. February 9, 16 and 18, 1916
7. **Yuma Sun**, March 17, 1916
8. Ibid.
9. **Weekly Holtville Tribune**, May 16, 1916, pg. 1
   Yuma Sun, March 17, 1916
10. **Weekly Holtville Tribune**, June 16, 1916, pg. 1
11. Ibid. July 27, 1916, pg. 1
    Yuma Sun, July 30, 1916
12. **Weekly Holtville Tribune**, July 31, 1916, pg. 5
    Also the Historical Files of the Automobile Club of Southern California, from a letter to the San Diego Historical Society date July 28, 1959 and signed by G. P. Parmele of the Auto Club. Located in the files at San Diego Historical Society vertical file #1. "It was constructed at Ogilby siding on the Southern Pacific Railroad. The sections were built of 3x8 inches, 3x10 inches and 3x12 inch planks, eight feet in width and laid crosswise on 4x6 stringers 12 feet long. Each 12 foot section was then hauled to the road site by teams of 16 mules."
13. **Weekly Holtville Tribune**, July 31, 1916
14. Ibid. August 8, 1916, pg. 1
15. Ibid. August 8, 1916, pg. 1

Travel on the Plank Road

1. The Wit of the Old Plan Road by Don Dedera: Off the Cuff Outdoor Arizona, August, 1976
2. **The Morning Sun** newspaper, March 22, 1924
3. Old Plank Road connected Yuma, El Centro by Dan Woods, **Tri-Valley Dispatch** newspaper, August 25 and 26, 1976
4. Goldrock Facts and Folktales by Iva L. Ouisinger, Monterey, CA 1973
5. Paving America's Sahara, Literary Digest, December 18, 1926
6. **El Centro News** newspaper, El Centro, CA, April 27, 1919, pg. 1
7. The Colorado by Frank Waters Rinehart, 1946
8. **The Morning Sun**, September 27, 1916, pg. 1
9. Ibid. September 28, 1916, 4:2
10. Ibid. October 14, 1916, pg. 3
    Holtville Weekly Tribune newspaper, October 10, 1916
11. From Dusty Trails to Super Highway by James D. Blick
    Brand Book Five San Diego Corral of Westerners, San Diego, CA, 1978
12. **The Morning Sun**, October 31, 1916, pg. 1
    Weekly Holtville Tribune, November 8, 1916
13. The Morning Star, January 9, 1917, 4:1
14. Ibid, May 23, 1917
15. Ibid. September 8, 1917
16. **The San Diego Tribune** newspaper, October 27, 1979, Vertical Fire-Road Misc. San Diego file #1 and #2, San Diego Historical Society Library and Manuscripts Collection, San Diego, CA

Plank Road Traffic Signs and Regulations

1. **The Yuma Morning Sun** newspaper, February 26, 1915
2. Ibid. May 26, 1915 1:5
3. Ibid. May 19, 1917
4. Ibid. February 26, 1925
5. Ibid.
6. Ibid. August 15, 1925

The Final Effort

1. From Dusty Trails to Super Highways by James D. Blick
2. **The Yuma Daily Sun** newspaper, October 24, 1954
3. Ibid.
   The Roadrunner #61, November 23, 1926
   Arizona Historical Society files.
   Conquering the Sand Hills by Ruth Borum
   Progressive Arizona, April, 1927
   Arizona Historical Society files
4. Ibid.
5. The Roadrunner #61
   Conquering the Sand Hills by Ruth Borum
6. Ibid.
7. Ibid.
8. Paving America's Sahara
   Literary Digest, December 18, 1926
   Yuma County Historical Society files.
9. The Roadrunner #61.